101 BLADESMITHING MISTAKES

TABLE OF CONTENTS

TABLE OF CONTENTS

INTRODUCTION

A Master bladesmith is an individual that has spent countless hours honing and crafting multiple skills to produce flawless knives and swords out of common metal and steel. When viewing one of these individuals in the process of creating a piece of work, it is easy to become captivated. You can quickly be pulled in and filled with excitement and eagerness to learn how to mold and manipulate steel the way they do. As exciting as it may appear to be, actually beginning the process yourself to becoming a bladesmith will be met with plenty of hiccups and letdowns.

This is not to discourage you. Even the most skilled Master bladesmiths have a list of mistakes they encounter and grew from as they transformed their techniques into perfect designs. While mistakes are expected, and necessary to truly learn the appropriate skills of bladesmith, some should just be avoided completely right from the beginning.

If you are someone who is just starting out in bladesmithing or interested in learning about the process, then you will find value in the pages that follow. This book lists the most common mistakes bladesmith's make when they are just starting out. From ignoring serious safety concerns and design flaws to just not understanding the importance of each step. These mistakes can leave you discouraged, hurt, and result in lacking the necessary skills needed to obtain the title of Master bladesmith.

But you don't have to suffer through these common mistakes. As you will learn, there are easy and simple ways to bypass these hard lessons. Whether you are looking to get into bladesmithing just for fun and use it as a side hobby or want to earn the title of Master bladesmith you will want to first begin by taking notes and not making these mistakes.

CHAPTER 1: WHAT IS BLADESMITHING?

Bladesmithing is an art form that has a history dating back thousands of years. Through this art, a bladesmith carefully designs, files, forges, hammers, and grinds unique swords, daggers, knives, and more.

Bladesmith Versus Blacksmith

A bladesmith and blacksmith are terms that are often interchanged. There are key and significant differences between the two. A bladesmith refers to an individual who makes knives, swords, or other blades. A blacksmith refers to an individual who can forge iron. The two can be the same. A blacksmith can forge whatever piece he or she wants from iron including blades. A bladesmith commonly uses steel to create blades but could also use iron. The steel used in bladesmithing is often a combination of iron, carbon, and sometimes additional alloys.

Stocking Versus Forging

There are two general techniques or methods used by a bladesmith to create a knife. The first is stock removal. Stock removal is the process of filing, grinding, or removing excess metal to form a blade. The second technique is forging. Forging adds the element of heat to the metal to shape it into the blade.

A skilled bladesmith knows how to stock and forge metal effortlessly. Individuals can begin to develop their skills as a bladesmith by learning to properly work a piece of steel. There are plenty of methods and techniques that can be used when doing stock removal and you must understand both techniques. Stock removal is a process that can be used alone to create a blade and it is also used a great deal when forging metal into a blade.

On the road to becoming a Master bladesmith, it is these two vital techniques that will ultimately determine whether they earn their respective titles or not.

What Does It Take to Become a Master Bladesmith?

Master bladesmith is the title given to those individuals who have spent years working in the bladesmithing field. There are many steps and requirements expected from someone looking to gain this respectable title. While you may just be a beginner, it is entirely possible for you to one day receive this recognition. To do so, you will need to check off the following requirements and pass specific tests that show your skill level.

1. Undertaking at least 2 years (preferably 3 or more) of apprenticeship.
2. After the apprenticeship, they can begin a series of tests, the first of which is to create a simple carbon steel 10-inch blade. This blade must be a razor-sharp knife and must also be able to:

- Slice through a free-hanging rope. The slice must be done with one single swipe and the cut must be made at least six inches from the bottom of the rope.
- Cut through two 2x4's. This is a time-consuming process that is done to test how the edge of the knife holds up when violently worked. If there is any deformation that occurs in this process, you fail.
- Test the blade on a finer material. It needs to be able to shave hair. This is to show that the blade has not become dull during the first two tests. This usually requires the person to have a patch of their arm hair to test on.
- Finally, the blade is tested for its strength. The tip of the blade is placed in a vise and is forced to bend at a 90-degree angle. This shows the individual's skills when it comes to heat treating the metal. If the heat treating was done properly, the tip should bend and spring back to its original shape without shattering or snapping.

3. If the hopeful Master bladesmith passes all the tests, they must then go through a round of grueling critics. They must present five blades to be judged by a jury of Master bladesmiths. They judge the knives based on their designs, construction, guard, balance, and proportions. They should present impeccable craftsmanship and be

flawless. If the jury approves of the designs, individuals are given the recognition of journeyman. Now you will be officially on your way to earning the title of Master bladesmith.

4. Once you have obtained the title of journeyman, you must put in two to three years of blacksmithing experience, and during that time, you must also have had an apprenticeship under your belt.

5. Once you have fulfilled the years of experience, you will need to take part in another series of Master bladesmith tests. During these tests the journeyman must:

    •   Make a Damascus blade that has been folded 300 times.
    •   The blade must contain a hidden tang.
    •   The blade will then be put through the same series of tests: cutting through a free-hanging rope, chop through two 2x4's, shave a patch of hair, and bend at a 90-degree angle.
    •   If the Damascus blade passes these tests without any damage, the five other blades must be brought in front of the jury of Masters. Each of these blades must be made from Damascus steel, one of which must be a quillon dagger. If all the blades pass the critiques of the master jury, then the individual is awarded the title of Master bladesmith.

As you can see, it might sound simple enough to acquire the Master bladesmith title, but once you begin forging your knife, you will begin to learn just how challenging it can be. It takes a great deal of patience and skill to become a Master bladesmith; there will be a great number of mess-ups and failed attempts. Luckily, some mishaps can be avoided when you pay particular attention and take care to avoid some of the most common mistakes. In the following pages, you will learn about the common mistakes a new bladesmith may across, including details about what not to do and what you should do instead.

CHAPTER 2: SAFETY MISTAKES

The first mistakes you want to avoid from the very beginning are the ones that can cause you physical harm or bring about any health issues. Bladesmithing can be dangerous when you do not take the appropriate steps to ensure your workspace is clean and that you are properly protected. When you first begin bladesmithing, keep in mind these safety mistakes you do not want to make.

1. **Not Wearing a Respirator When Grinding or Sanding**

When you are forging your work, all that sot, sand, and fragments can be easily inhaled. These particles can cause severe damage to the nasal cavity and when they make their way to the lungs, it can cause permanent damage. You must wear a respirator or protective mask when you are grinding and sanding to avoid serious health issues. Since it can take years for the negative effects of breathing in these materials to occur, many never know the dangers of breathing in the constant dust and debris until it is too late.

2. **Not Having Your Shop Set up in a Space That Allows for Proper Ventilation**

Piggybacking on wearing a respirator, you want to ensure your workstation is properly ventilated. When you are forging, you will be burning fuel which increases the carbon monoxide in the area. This can cause fatal effects. You want to ensure that you have proper ventilation in the room you're working in. Have fans blowing to move air around in the room and have windows or ducts in place that will help to remove the toxic fumes and gasses from the area. Have fans blowing to take away any smoke produced from the forging system drafts.

3. Not Wearing the Appropriate Eye Protection

Eye protection for some reason is the first overlooked safety precaution many beginners believe they can ignore. When you are forging, welding, or hammering hot pieces of metal, the sparks that are produced can fly straight into the eyes and cause serious damage. You must wear some kind of eye protection. Find a quality pair of safety glasses or goggles, and always wear them when you begin working.

4. Not Thinking to Protect Your Ears

When working with metal, hammering, grinding, and sanding can make up for a lot of noise. This noise can damage the ears and cause hearing issues in the future. When you are hammering, you want to ensure you are wearing some type of ear protection, such as headphones or earplugs that will reduce the loud noise you are being exposed to.

5. Not Cleaning Your Machines After Each Use

This is especially dangerous when it comes to the forge you use. You want to ensure that debris and particles are removed from the forge after you use it. Neglecting to skip this vital step can lead to internal fires sparking up when you are not around, which can lead to total devastation.

6. Working When You Are Tired

The more tired you are, the more likely you are to make mistakes and cause injury. Do not force yourself to continue to work on a project when you know you are getting fatigued. The same applies to being physically tired as it does to being mentally tired.

7. Being Afraid of Injury

Many beginners overthink how to properly use the tools when bladesmithing. This causes them to approach the tools in a rigid and tense manner. They focus too much on the injury that can occur instead of how to properly hold and use the tools to avoid injury. If you are afraid of exerting the right amount of energy when you are hammering a piece, it is more likely that you will not only make a mistake but that you will also seriously injure yourself. If you are too rigid when using the tools, you lack the control you need to avoid injury. Have confidence in your abilities to use the tools and you will avoid internal and external injuries.

8. Wearing the Wrong Clothing When in Your Workspace

When you are crafting a knife, you will encounter many flying sparks. You cannot predict where these sparks will fly and oftentimes, they will fly right towards you. If you wear short sleeves T-shirts and are not wearing a welding apron, you are setting yourself up for painful and dangerous conditions.

When you go into work, you will want to wear long sleeves to protect your arms from any flying sparks. More importantly, you want to be wearing a long, heat protecting welding apron. The apron should be made from a leather material that won't catch flames when sparks land on it.

9. Not Listening to Your Body

Stress injuries are a common discomfort for a bladesmith. The repetitive hammering can cause stress injuries over time. To avoid long-term side effects from working as a bladesmith, it is important to always use proper techniques when hammering. Some tips to avoid stress injury:

- You want to ensure that you are holding your hammer at the appropriate height.
- Always take note of pain you feel in your hands, shoulders, or elbows.
- When holding the hammer, you want to be sure to wrap your thumb around the handle of the

hammer. Many hold the hammer with their thumb simply resting on the handle. However, wrapping the thumb around the handle not only gives you more control but it also reduces stress on the wrists and elbows.

- Plan to take breaks when you are doing a lot of hammering. Take a few minutes to stretch to avoid injuries.

10. Not Having a First Aid Kit and a Burn Kit in Your Shop or Work Area

Burns, minor cuts, and scratches come with the territory of bladesmithing. The biggest mistake you can make is thinking these types of injuries will not occur. You want to be prepared for them by having a first aid kit easily accessible in your work area. This first aid kit should include ointments or medicines to treat burns and cuts to provide you instant relief and to properly clean areas that get scratched.

11. Not Paying Attention to What Is Around or near Your Forge

You want to pay particular attention to the flame of your forge. While you may be close to the forge when you are working, you also want to be aware of what is around the forge. Keep items that are dry and flammable away. Keep the flame of the forge contained and be aware when it might be getting out of control. Never leave the forge unattended.

12. Leaving Hot Metal Just Laying Around

Many beginners fail to have a designated space where they will place hot metal. This, of course, leads to many burns and accidents that can be easily avoided. Whether you work alone or with others, a common mistake bladesmiths make is not having a designated place for the hot material you use. When you randomly place hot metal wherever there is space, you can easily forget that it is hot, or others may not be aware that it's hot, and this will result in burn injuries. You want to set up an established space where you will always place the hot metals or tools you work with. That way, if you see any metals or tools in this space, they are most likely hot and should be tested before handling. You should also place signs up that will let others know that the items in that area should not be touched unless they are wearing the proper safety gear.

CHAPTER 3: MOST COMMON BEGINNER MISTAKES

These mistakes are often due to ill planning or poor mindset. Most often these mistakes are made due to the excitement of wanting to jump right into completing the first knife. While a little excitement is good to have, when it gets in the way of properly executing the bladesmith process, it can become a major setback.

1. **Using a 'Super Steel' When Making Your First Few Knives**

The most common mistake for beginner bladesmiths is choosing the wrong steel to learn on. When you first begin bladesmithing, you want to use a steel that is easy to learn with. Super steels require specific heating and cooling temperatures and times. While these are things you need to eventually know, what you first need to learn is the proper techniques. A 1084 steel will allow you to learn as you go. It is easier to manipulate and is considered the easiest steel to work with. It is also a great deal cheaper, so when you make a garbage-worthy design on your first few goes, it won't hurt your budget as much as ruining higher-priced steel.

This is not to say the quality of the steel is less desirable. 1084 steel is much easier to work with when learning about proper heat treatment. It is also cheaper, so learning on this type of steel makes sense both in developing the skills you need and saving money as you learn.

2. **Thinking You Need to Have the Same Tools and Fancy Equipment You See the More Skilled and Advanced Bladesmiths Using**

When you are first starting in bladesmithing, the only tools you need to have are:

- File or sandpaper
- Drill press or hand drill
- Hacksaw

These tools are necessary to learn how to file and grind a piece of metal. This is the first step and set of skills you will need to develop when bladesmithing. You don't need to run out and buy or make a forge, or any of the other fancy gadgets. Just the basic simple tools will allow you to get started in bladesmithing. Being able to properly grind and sand a blade to razor-sharp perfection is a vital skill you will need when you move onto learning to forge or heat treat the metal.

Having the best tools will not only make you a good bladesmith but knowing how to use the tools and practicing with them will make you a great bladesmith. You need to have the basic tools that will allow you to learn the proper methods and techniques. Until you get those down, you don't want to waste your money on any fancy and expensive tools.

3. Simply Trying to Hit the Piece Again When You Miss on the First Strike

When you are first learning to strike the metal, there are plenty of mistakes that will be made. One of the most common mistakes is not learning from a mistake you just made, such as missing the piece you are trying to strike. If you attempt to hit the piece and you miss, this is a good time to rethink your position. Simply trying to hit the piece again will often result in a second miss. If you miss the piece when you go to strike, this is due to the piece not being positioned correctly on the anvil. You want to stop trying to hit the piece. Instead, reposition yourself or the piece, and then strike in the right direction.

If during this process you have made contact with the piece, but it has caused damage, you will need to repair the damage first. Do not just reposition yourself or the piece and continue to strike, this will cause further damage that you may not be able to fix later on. Know what to look for when your hammer strikes. The best bladesmith has developed their skills by paying close attention to how the hammer affects the steel when it makes contact and they know when it is causing imperfections in the piece.

4. Not Having the Tools Already Laid out When You Begin

You will notice that planning is a vital step in a bladesmith's process. Before you even begin a project, you want to ensure you have the right tools available for each step, What size tongs will you need? Do you have all the sandpapers you need? You don't want to begin a project and then realize halfway through that you have the wrong size tongs, you don't have the right grit sandpaper, you've misplaced your hammer, you don't have the right drill size, and so on. Take the time to plan and you will avoid several mistakes along the way.

5. Not Planning on What Needs to Be Done as You Move Through Each Step of the Forging Process

When it comes time to forge your piece, you want to be sure you have carefully thought about all the steps you need to take and what tools you will need. Before you begin a project, ensure you have all the tools neatly organized on a workbench close to your anvil. You want to avoid leaving your hot metal sitting around while you run to get the right tools.

What you also need to think about when forging is how you will go about striking and shaping the piece. You want to limit the number of times you heat the metal; therefore, knowing exactly what you will do when you remove it from the heat is crucial. You will need to have patience when you are waiting for your metal to reach the appropriate temperature, therefore, giving you enough time to think about what you need to do next. How do you hold the piece? What shape are you trying to achieve? A Master bladesmith is always thinking

ahead so that they know exactly what needs to be exe-
cuted to achieve the desired design.

6. Thinking the Hammering Hand/Arm Will Be the One Doing All the Work

As you set out to work on your first piece, you will
quickly realize that the hand doing the holding gets just
as fatigued as the hand swinging the hammer. The swinging arm takes on a lot of work when you are working
your metal, and the stress from the pounding can cause
some serious damage. But this doesn't mean the other
hand is just relaxing. The holding hand takes on a lot of
stress too. When you are hitting the work, your holding
hand needs to be able to keep the metal at a specific angle
and hold it steady as you hit. Many beginners fail to realize how much of a challenge this can be.

7. Using Little, Ineffective Blows When Shaping Your Piece

Lacking confidence, knowledge, or recognizing the type
of blow that will affect how the piece is shaped is a common mistake you need to address and overcome quickly.
Using dinky blows when you are forging might be acceptable at some point in the process but should not be
used each time you strike the piece. Confidently raise the
hammer above your head and give the work an effective
strike to produce the result you are trying to achieve.
Dinky light blows waste the time you have to work on a
piece and can require you to constantly heat the metal

over and over. Your goal should be to strike with accuracy to limit the number of times you need to heat the metal. Many beginners fall into the routine of using light strikes because they fear hitting the piece wrong and ruining their work. Learning how to use the hammer to shape, strengthen, and mold the metal takes time to fully develop. It will take a great deal longer if you don't put more effort into it.

8. Not Knowing the Proper Temperature Your Workspace Needs to Be so the Epoxy Used Goes on Smoothly

The warmer your workspace, the faster the curing time. This can result in having less time to work the epoxy. If on the other hand, your workspace is too cold, the epoxy may be too thick and clumpy to work with. Even if you are able to work with the epoxy when the temperature is cooler, it can take days for the epoxy to cure and sometimes even after that amount of time it may still remain tacky or sticky. Ideally, you want your workspace temperature to be a few degrees above the manufacturer labeled cure temperature. This is typically around 73 degrees Fahrenheit. Try to keep your workspace at around 75 to 80 degrees. It is also crucial to point out that as you are waiting for the epoxy to cure, or for at least the first 24 hours after it has been applied, you want the temperature to remain consistent. If there is a sudden drop in temperature, you may find your work has white wavy streaks or lines going through it.

9. Jumping Right into the Process

Once you have that slab of steel in your hand and your file close by, it is easy to immediately start grinding away. You begin to remove the steel and shape your slab, and then you get a decent design that could resemble a blade. Except you haven't considered the appropriate size the blade should be for its intended purpose or how big the handle will have to be to make it possible to use.

Many beginners jump right into grinding and sanding their steel without giving any thought to the purpose or design of the finished blade.

10. Rushing Through the Process

The way to develop your skills as a bladesmith is to take the time and actually see how each hit affects the metal. Only by taking your time will you notice these subtle differences. Take your time with each step of the process and do not try to cut corners to reduce the time it takes to complete a step. Trying to skip out on what you think doesn't make a difference can result in a poorly finished product and discouragement. Don't rush through the process; your work will show how little time you actually took to carefully perform any of the steps.

11. Not Understanding the Basics

The first step you should take if you really want to learn how to become a bladesmith is to learn from an experienced bladesmith in your area. This will allow you to learn the basic set of skills necessary before you run out and buy the necessary equipment. Learning is a part of the process. While there are plenty of books, videos, and articles online that you can learn from, actually seeing and being there in person as someone explains the process will give you far greater understanding and appreciation for the craft than you could gain from online resources. After you have done your time working with someone else, you can begin to acquire the necessary tools to supply your own home shop.

12. Trying to Make a Large Complex Knife for Your First Project

It can be enticing to want to jump right into sword making and creating detailed, one-of-a-kind pieces, but these types of projects require a great deal of understanding of the bladesmith process. You need to go through the motions and fully understand how much work is required for each piece prior to undertaking something as complex as a sword. You will find it takes more time, effort, and energy just to create a small piece. By beginning with a small piece, you will better refine your skills, which will need to be fully developed before moving onto more complex knife making projects.

You want to keep it simple when you are first starting out. Making mistakes on a smaller piece is much more bearable than when you make the same mistake on a larger piece. Start with a small or medium-sized knife.

Once you have perfected these smaller projects, then move on to daggers and bowie knives.

13. Being Too Nervous During the Process and Messing Up

Nerves will always get the best of you. If you constantly let your nerves control your motion when bladesmithing, you will either constantly make mistakes or never move beyond the basic skills. Messing up is how you learn what not to do. Don't be afraid of making a mistake. Instead, focus on how you are holding your piece, the tools you are using, and the effect you are aiming for. Focusing on the mistakes you might make is a sure way to guarantee you will make a mistake.

14. Not Creating a Design Before You Begin

Many beginners do not even think to draw up a diagram of the blade they intend to make before they get started, or they just have a simple sketch to work from. Before you begin the bladesmith process, you should take your time to draw out what you want to create. This should be done on graph paper and should be close in scale to the size you want the piece to be. Have the correct measurements planned out for the working blade, as well as the materials you will be using for the handle. Once you have your diagram drawn out, you can transfer this to your slab of steel. Trace the outline with a permanent marker, which will allow you to know exactly how long and wide the blade needs to be, as well as how long the tang of the handle has to be. This is a crucial step in the process that can make all the difference in the final look of the piece.

15. Thinking the Work Is Good Enough Just Because You Want to Move On

When you are just starting out, mistakes should be welcomed. But instead of just ignoring them and saying a piece you are working on is "good enough," you should make the effort to learn from the mistakes. Never be done with a piece unless you are completely satisfied with the end product. There are ways you can correct some common imperfections in your work such as making adjustments to the scale or sometimes starting all over again. Bladesmithing takes practice. This cannot be stressed enough; you need to be willing and eager to start projects over to perfect your craft and improve your skillset.

16. Not Finding the Center on the Tang

The tang is the backbone of the knife and its core responsibility is to hold the knife together. If you do not take the time to find the center of this tang, the whole balance of your knife can be thrown off. This doesn't just come at a cost of balance though. When the center of the tang is not accurately identified, the use of the knife can be compromised. Users will find that the blade to handle balance makes it difficult to maneuver the knife.

17. Having Sharp Inside Corners

Sharp inside corners create stress points on the blade. These can increase the risk of the blade being easily damaged and ruined. To avoid these sharp corners, you want to use sandpaper to smooth out the edges slightly. Use epoxy to fill in holes made and to provide more support.

CHAPTER 4: WHAT TO AVOID WHEN MAKING YOUR FIRST KNIFE

Now that you have an idea of the basic mistakes to avoid, you will find yourself encountering many more as you actually begin making your blade. There are a number of steps you need to take when creating a blade and each step poses a number of mistake opportunities. While some mistakes will serve as a learning experience, there are some that you can avoid by following the advice offered in this chapter.

1. Not Having a Designated Place for Your Hammer While Forging

You have a limited amount of time to work your metal when it comes out of the fire. You don't want to spend any of that time looking for where you set down your hammer. Have a designated place where you will place your hammer while heating your work. Anywhere near or hanging on the anvil is ideal, but you can also simply keep the hammer in your hand or hang it from your work belt. Just be sure to get into the habit of keeping it in the same place every time you go to forge.

2. Changing Techniques Too Often

There are plenty of techniques you can use during the knife making process. Some of these techniques can be used no matter what type of knife you are working on, while others should only be used for specific types of knives. Not understanding which techniques are best for the knife you are making can result in a colossal mess. When you are just getting started in bladesmithing, you want to learn one technique at a time. Do not try to change things up in the middle of the process. Changing your techniques too often, without having mastered them individually, will lead to a confusing design that is off-balanced, poorly finished, and useless.

Learn and master one technique at a time. As you learn one technique, you can then begin to add on additional ones to see how they can be used together; but resist the urge to master everything at once.

3. Giving a Shearing Blow When Placing Notches in the Knife

When adding notches or choil to your knife design, many beginners think they need to exert a great deal of force to accentuate this detail. This often causes the metal to get crushed or become smashed as opposed to creating a notch. When working your metal to create a notch, you want to strike the piece with enough force to cause an indentation in the metal. The amount of effort in the strike is only part of the equation when creating notches in the design. What needs to be more importantly considered is how you hold the work on the anvil and where your hammer actually connects with the piece. To create a notch, you should first hold the work halfway off the anvil and have the hammer strike in between where the piece sits on the anvil and where it hangs off. This will form a notch.

4. Not Knowing How to Finish the Choil End of the Blade Before Gluing

The knife choil indicates where the cutting edge of the blade ends, and the blade tang begins. It is a little notch in the knife that helps determine where the blade needs to be sharpened. If you neglect to finish the key feature, then you run into the problem of not knowing where the handle should start, which then results in having a larger or smaller blade than you anticipated creating. It will also make it more challenging to sharpen the blade when finished and can make using the knife a challenge when there is not an area to place the index finger needed to guide the stroke of the knife.

5. Not Using Tape on the Blade When You Disassemble the Knife

When disassembling a knife, either to fix the design or add different elements, you always need to think of the blade. Using any type of tool near the blade can accidentally scratch it. To give the blade protection from the tools being used, you want to tape the blade. This is especially important when you run into trying to remove a stubborn screw that won't easily dislodge. The blade can be scratched from any slip of the tool and this can permanently damage the blade.

6. Changing Designs Half Way Through the Process

The change in design mistakes seems to be the common way beginners cover up a total screw up in the bladesmithing process. If you set out to create a specific knife design, do not change things up when you make a mistake, even if it just seems easier to change the whole design instead of looking closely at the work and making the appropriate corrections.

7. Not Matching up the Plunge Lines

Plunge lines let you know where you are supposed to start grinding the blade. There should be a line on each side of the blade which will often be blended into the actual blade or flattened to create a 90-degree angle. If you do not accurately line up your plunge lines on the corresponding side, your blade angles can become imbalanced. You need to take the time to properly align and triple-check that the plunge lines are matching up before and as you are grinding the blade.

8. Have a Blade That Is Too Thick or Too Wide

The thickness of a blade will often suggest how sharp the blade can be made. Although thick and wide blades can be sharpened to a razor-sharp edge, this usually requires a great deal of stock removal. This can take a lot of excess energy and time. When you are first starting out as a bladesmith, it is best to learn how to create a blade by using a thinner blade.

9. Having a Handle That Is Too Large

Knives need to have a nice balance between the blade and the handle. If you have a short blade and a large handle, the knife may not be able to be used properly. This imbalance can make handling and maneuvering the knife challenging to the user. The handle length depends on the type of blade you are creating; if it is too large, your design will look flawed.

10. The Resulting Knife Looks Stiff

A stiff knife is one that has straight edges. There are almost no curves in the blade at all. While this can be the look you are going for and can be a great way to start learning how to create a flat blade, it doesn't result in a very appealing-looking knife. Adding some curvature to your design will give the blade more character and design appeal.

11. Applying a Mirror Polish over a Bad Sanding Job

It there is one thing that will make all the sanding imperfections stand out, it's using a mirror polish to finish your work. A mirror polish is a great way to finish off your blade, but it won't look so pretty when the blade is full of scratches or is not evenly sanded. To avoid a terrible mirror polish, you want to really take your time to slowly and fully sand the work, working through each of the sanding grits from roughest to finest.

CHAPTER 5: IMPROPER USE
OF TOOLS

A bladesmith doesn't need many tools, though there are a variety of the same tools they can use such as different hammers and tongs. The most basic tools include a forge, anvil, various hammers, vise, and tongs. When used properly, you can easily complete a decent knife, even as a beginner. When used improperly, you not only risk ruining your work, but you can cause yourself physical harm. In this chapter, you will learn about the number of ways not to use your bladesmith tools and what to do instead.

1. Keeping the Anvil at the Wrong Height

Anvil height has an important role in the bladesmith process. Not only should it be at a comfortable height, but it needs to be at a height that allows for the most effective placement of the piece you are working with. When working at the anvil, you should never be hunched over while hammering or holding the piece. In the long run, working in this manner will cause excessive strain and damage that will result in a great deal of pain. Your back should be straight to take the strain off your back and you should be able to hold the piece you are working on comfortably at the appropriate angle. If you find yourself bending over the anvil while you are hammering, you need to raise it.

2. Bobbing the Head and Torso When Hammering

Your body should not be all over the place when you are hammering. When your head or torso is bobbing excessively, you do not have much control over what you are doing. Though you do want to keep your body relaxed when hammering, you want to also maintain control over it. When you position yourself at your anvil, keep your feet about hip-width apart and the knees should be slightly bent. The back should remain straight with just a slight curve. When you are hammering, it is not your torso or head that should be moving up; your arm should be raised above the head to allow for more precision.

3. Not Keeping the Stock or Work Parallel to the Anvil When Strengthening

If you are holding the work straight out in front of you while you hammer, there is a good chance you are not hitting your target area. When you are striking the piece, you want to position your body so the work remains parallel to the anvil so that you can see exactly where you are striking. This is especially important when you are strengthening the metal, where you need to add more volume in certain areas of the blade and thin out others to create the blade.

4. When You Are Striking for Someone Else You Use a Regular Handheld Hammer

When you apprentice at a bladesmith shop, there will often be times that you will need to do the striking for someone as they hold and move the metal under the hammer. If a second person is needed during the forging process, this is typically because the piece is a heavy-duty blade. When you are striking a piece like this for another person, a standard handheld hammer may not supply ample force. A sledgehammer is more effective, and its purpose is to be used for heavy-duty striking.

5. Holding the Haft Too Tightly

Haft is a term for the handle that you might hear used more often when creating swords or daggers. Just as with any other steps in the bladesmithing process, you do not want to have a death grip on the haft. Your grip should be relaxed and loose when you are holding the haft but firm enough, so you maintain control over the piece.

6. Being Stiff While Hammering Your Piece

When you hammer, you should never be stiff. You want your stance to be more relaxed as if you were going to pitch a game of baseball. Tensing the body while you are hammering not only increases the amount of stress you are placing on your body but also takes away the control you have over the hammer.

7. Trying to Cut Corners by Pushing or Sliding the Hammer While Hitting

While there is some logic behind this mistake, it is still a big mistake that will reduce the strength of your final piece and can ruin it in the process. Beginners believe that hitting the piece with a sliding or pushing motion will help lengthen or draw out the piece more quickly. This can cause unwanted, and unrepairable, splits and cracks in your work. When hammering, you always want to hit the piece straight on. This will naturally cause the piece to draw on its own with minimal damage while working.

8. There Are Gaps, Spaces, or Light Visible Between the Piece You Are Working on and the Tong Jaws

Not properly fitting the tongs around the piece can be a very dangerous mistake. When you are using the tongs, you want to ensure that there are no spaces or gaps between the piece and the inner jaw contact. You want the piece to be held firmly in place to avoid any slippage when working on your piece.

9. When Using the Horn of the Anvil, You Bend and Hit Against the Horn

When you are hitting your piece against the horns, you are striking your piece with leverage blows. This won't help you shape or bend the metal in the way you desire.

When you are holding your piece and using the horns of the anvil, you want to strike just beyond the horn.

10. You Do Not Move Your Piece Under the Hammer

The hand doing all the hammering is not the only one that needs to be moving and working. When you are striking your piece, you need to know when to move the work so that the hammer hits it at the perfect point of contact. Instead of trying to re-adjust where the hammer will hit your piece, you need to adjust the piece so that the hammer will hit where it is intended to hit.

11. Trying to Work Against Gravity

When you are trying to shape your work, it is a natural reaction to hit the work in the direction needed to create the desired shape. In other words, you will hit the piece upwards or at an angle instead of turning the piece. This not only will cause inaccurate blows, but you are not allowing gravity to work with you. Striking your piece upwards or sideways will result in you having to use more energy and take up more time. Instead, you should always turn the piece, so you are hammering downwards. This means you are always keeping your swinging motion in a comfortable position and you have gravity working with you. You will not need to use as much force or energy to get the piece to bend in the way you want.

12. Using the Same Belt for Far Too Long

Using an old, worn down or tattered belt for too long won't help you save money, time, or help you become a better bladesmith. You need to know when it is time to put a new belt on and do so when the time comes. Using the same belt for an extended period of time will result in a piece that is unpolished, and not very well crafted. Changing the belts on your grinders is an easy and quick process; much quicker than having to continuously sand and grind a piece over and over using a worn-out belt.

CHAPTER 6: USING THE WRONG
TOOLS, MATERIALS, OR METALS

When creating a knife, there is a specific reason why certain tools or materials are used during the process. Many beginners think they can just use whatever they have available, but this is a hard lesson you want to avoid having to learn through personal experience. This chapter covers the mistakes many beginners make when they do not use the appropriate tools or materials when creating their blade.

1. Using the Wrong Tongs

There is no one size fits all when it comes to the tongs you will need through your bladesmithing adventures. If you are just starting out, you can get by with one tong, since you will primarily be making the same size or designed knife. But, at some point, you will need to expand your tong collection. Not using the right tongs when holding your work can quickly result in a terrible accident. The work can easily slip from the tongs, especially when hammering.

2. Using Acetone to Clean a Kydex Sheath

Using acetone on a Kydex sheath will result in a white film that ruins the look of the sheath. The best way to clean a Kydex sheath is old-fashioned water. You then want to pat it dry and let it sit for a little, then use compressed air to blow away any excess water build-up and let it air dry the rest of the way. Acetone can leave a white filmy residue ruining the look of the sheath. On top of that, it can cause the material to turn rock hard.

3. Using Goop Quench

Quench is used when heat treating your work. There are many different varieties of quench you can use for this process, but your best bet is to use pure vegetable or canola oil. Goop quench is a combination of used cooking grease, typically bacon grease, wax, and motor oil. This combination may be useful in heat treating. You have to take into consideration the excess particles that will be found in the used grease and wax that can damage the blade. It is a better idea to use clean oil for quenching.

4. Using a Dremel When Making a Bowie Knife out of Unannealed 440C Steel

A bowie knife is a large, very versatile and heavy-duty knife. Using a Dremel with cut-off wheels will not only prevent you from making accurate cuts to the 440C steel when cutting out the design, but it will also slow you down. The Dremel wheels will easily wear out. A hacksaw is a much faster and effective way to start with a bowie knife.

5. Thinking You Can Save Money and Get More Use out of Your Sandpaper or Belts

There are other ways you can save money. Using sandpaper and belts way past their abilities does not save you money and it wastes a lot of your time. When the sandpaper is used, the grit of the surface diminishes, which affects the ability of the sandpaper to properly sharpen the edge of the blade. If you continue to use this dull paper, you will end up with a dull knife.

6. Using Aluminum Oxide After You Have Developed the Necessary Grinding Skills

Aluminum oxide belts are ideal to learn on. They offer a wide range of grit and are fairly inexpensive. They can also help you learn how to better contour your pieces and you can produce some decent knives when using aluminum oxides. That being said, once you have obtained the skills necessary to skillfully use an aluminum oxide grinder, you'll want to upgrade to something that offers more precision and high-end quality. Ceramic grinders offer a wide grit range, but they also self-sharpen. This means you won't have to replace your grinder as frequently as you would with aluminum oxide grinders. They are more expensive but well worth the investment if you want your bladesmith craft to be taken seriously. These grinders provide you with a well-polished finish and can be used to make precise cuts. This is one of the tools you will want to save up for and wisely invest in as you develop your bladesmith skills.

7. Buying a Thick Slab of Steel

When it comes to the thickness of the steel you use, you should aim for a thickness of ⅛ to ⅜ of an inch. This will require you to remove an excessive amount of metal when you begin your project, so it will cut back the time spent on the first few steps of the process. When you start with a thick slab of bar steel, you will need to take the time to grind off the metal. This can result in a blade being too thick or wide and increases the chance of mistakes being made.

8. Using Permanent Markers on Anything but Metal

A permanent marker is acceptable to use on the metal you are working with because it can easily be sanded off. Using a permanent marker on any other material like the leather of a sheath will result in unsightly lines that you won't be able to easily remove. When you need to place markings or lines on your other material, it is best to stick with a washable marker so you can easily wipe away the lines when needed.

9. Using Mystery Metal When Making Any Types of Knives

Each metal has its own specific set of properties and each will need to be treated and processed according to those properties. If you just use a combination of steel, metals, or whatever is laying around, you run the risk of not using or treating it properly. The result can be a mismatch of mistakes and unfavorable flaws in the final design.

Instead of using mystery metal, always use metal where you know the properties and can, therefore, treat it appropriately. If you want to make a quality knife, even as a beginner, it is imperative that you use a known metal instead of a mash-up of different metals. This is not to say that as you gain a better understanding of the metal properties you can't use scrap metal to forge a knife, but you need to have a clear understanding of the metal composition before you begin working with it.

CHAPTER 7: SKIPPING STEPS
IN THE PROCESS

There is no way around the steps you need to take in order to produce a quality blade. Skipping steps as you go or trying to cut corners to save you time, money, or effort will show in the work you produce. If you want to take the bladesmithing craft seriously, then avoid making these mistakes as you create your first knives.

1. Not Having a Design and Plan Before You Begin

The first step you should accomplish when you begin a new blade is to have a design drawn out. Before you begin gathering steel, the tools, or anything else, you need to have a design. Without a design, you will have no idea what you need to do with the metal or the tools. Many beginners think they can skip this step because they just have an idea in their head they want to try. This approach is a huge mistake. When you just go off of what you envision in your head, you have no clue what measurements or lines you want to create in the final product. When something does go wrong in the process, you have nothing to refer back to, to try to correct the mistake from happening in the future.

Take the time to draw out your design. This will not only supply you with a guide to follow but will allow you to take notes and keep track of what may or may not have worked, or what you might want to try differently next time. This design will also allow you to create a step by step plan as to what exactly you need to do during each step in the process. From how to shape the blade to heat treating to sanding and grinding to finishing the fit and handle. Design and plan the whole process before you begin.

2. Not Using the Basic Tools When You Make Your First Couple of Knives

When you first begin bladesmithing, do not jump right into using machine grinders and sanders. You need to understand how to properly angle and hold the knife when sharpening or removing metal. This is best done when you use the basic tools by hand. Take the time to properly learn how to file the metal and sand it using different sandpaper. You will learn how to avoid causing unwanted scratches in the metal and how to properly sharpen the blade.

3. Not Stropping the Blade When You Are Done Sharpening

Whether you must strop or not after sharpening your blade is a matter for debate. Some skilled bladesmiths say you shouldn't need to strop for a sharper blade if you use the proper fine-grit sandpaper when you sharpen your blade. Others say stropping makes a difference in how well it cuts through fine material like hair or paper.

In other cases, there is one clear benefit to stropping. It gives your blade a refined and polished look. For beginners, stropping can help refine your sharpening skills, but most agree that this step in the sharpening process should not be relied on to get the desired sharpness of the blade.

4. Not Having Holes in Place When You Heat Treat

Some believe that drilling holes after heat treating will allow them to drill more accurate holes. This can result in a number of mistakes. First, trying to drill holes after the metal has been treated results in a much harder metal, making it quite difficult to drill holes in the first place. Second, holes should be drilled when you have a nice flat surface. And third, when you drill your holes before heat treating, you want to make them just slightly bigger than the pins you will use. This will allow the proxy to fall in around the pin and set more securely.

5. Not Spacing and Aligning Your Rivets

Rivets are used to help hold the handle material in place on the tang. If you do not properly space these rivets, you run the risk of the handle not staying in place when being used. You also run the risk of the handle material splitting away from the tang in the areas that do not have rivets close by. When creating your initial design, you should already have a place where the rivets will go and an idea of how far apart they need to be spaced.

6. Not Completing Your Grind to Achieve a Flat Scale or Tang

Your scale and tang need to be perfectly flat if you want a balanced blade. Often times, beginners will skip grinding their work because they believe it is flat enough once they are done with the forging. This is a costly mistake that can ruin all your effort up until now. You will want to take your time to grind down your work after forging and be constantly checking the work to ensure it is maintaining a flat surface.

7. Not Using Dry Wooden Scales

Epoxy won't hold well to surface that is dirty or damp. You always want to make sure you have properly and thoroughly cleaned the surface area before applying your epoxy. Failing to ensure the surface or wooden scales are thoroughly dried before applying the epoxy will result in a blade that will quickly fall apart after just a few uses.

8. Using a Finer Grit When You Begin Sanding

There's a reason why there are different grit sizes. Each size allows for more precision and removal of unwanted marks and metal. Even if you think you can start with a lower grit size, you want to at least give a few passes of rough grit to ensure that the metal is clear of hammer marks or scratches. You also want to ensure that you go in a progression from the roughest grit to the finest. This process will make all the difference when it comes to the sharpness of the blade as well as the longevity of it.

9. Not Fully Finishing or Properly Fitting Your Knife When You Complete a Piece

Many beginners get into the habit of only halfway finishing a knife. Then, when they get to a point where they are capable of producing a decent blade, they ruin it by not properly fitting the handle. Even if the first few blades you make are subpar at best, go through the whole bladesmith process from start to finish. This will help you fully develop all the skills necessary to complete a quality blade.

10. Using Parts You Are Not Happy with When Constructing Your Knife

As a beginner, do not get into the habit of just finishing a piece just to say you have finished it. Whether the knife came out exactly as you designed it, or not, never throw pieces together that you are not happy with. If there is something you don't like about the blade, take the time to go back and fix it or start all over so you are more pleased with the final product. Get into the habit of creating work you are proud of right from the beginning and you will perfect your craft.

11. Having Uneven Grinds on the Blade

If you do not take the time to evenly grind the blade, you will run the risk of warping it when you go into the heat treating phase. Uneven grinds are easy to avoid but, as with all the steps, it just requires a bit more patience.

CHAPTER 8: MISTAKES WITH HEATING AND FORGING

There are plenty of mistakes that can be made when heating and forging your work. As a beginner, some of these lessons are learned through trial and error, even if you have the best understanding of the process. Other mistakes can be avoided altogether, like the ones mentioned in this chapter.

1. Building a Low and Flat Fire

This common mistake comes from the thinking that building a low and flat fire will result in less fuel being used. This is done to help preserve fuel, but in the process, you never build up the fire so it will properly heat the metal to its required temperature. You can end up just barely heating the metal even though you could leave it in all day. The proper fire needs to be built up so that the steel can heat properly.

2. Letting the Fire Thin Out

Along the same lines of building the proper fire, you never want to let your fire dwindle down. It takes time for the temperature to come back up, and at that time, you can end up ruining your piece. Always be sure to keep a careful eye on the fire and don't be stingy with the fuel. If you see it dying out, build it back up.

3. Taking the Metal out of the Fire Before It Has Reached a Lemon Yellow Color

Many beginners become impatient waiting for the metal to reach the right temperature during the forging process. Often, they remove the work when it is a reddish color, get a few hits in, and then have to return it back to the flame. If you remain patient and you wait for the work to burn a nice yellow color, it will be easier to shape your piece and you will be able to work on it slightly longer. This will help you use your time more wisely and reduce the time your metal spends in the flame.

4. Not Using Your Common Sense When Picking up a Hot Iron

This really shouldn't have to be discussed, but surprisingly, it is a huge and painful mistake a beginner tends to make. Those not used to bladesmithing using a forge have the misconception that just because the metal is not glowing red, that it is cool enough to touch. Just because one end of the work has not been kept in the flame when forging does not mean it is not hot. Always use gloves when handling metal around the forge or picking the metal up with tongs.

5. Not Knowing How to Identify the Welding and Coal Temperatures

You will hear a number of bladesmiths reference the color of the steel, as has been done in this book, in order to gauge when the steel is ready to be worked on. There are varying opinions about the accuracy of this method for gauging temperatures. While you can invest in a high-temperature thermometer, the color method gives you a general temperature range.

When the work begins to turn red, it has typically reached around 1100 degrees F. Once it changes to a more cherry red, it is around 1300-1400 degrees F. When it is glowing orange, it is between 1600-1700 degrees F. A yellow-colored steel will be around 1900 degrees F and a white-colored steel is usually around 2200 degrees F.

The main debate about this temperature gauging process has to do with the amount of light in your workspace, which can cause the color of the metal to appear differently as it is, and of course, the type of steel you are heating. While steel can be forged, once it has reached the color red, it will cool rather quickly, and you won't have much time to work. It is also not as malleable when it has reached the color red. White is a good indication that the steel is too hot and that you have probably burned or damaged the steel. You want to aim between an orange and yellow color when shaping your metal.

6. Warping the Edge of the Blade When Heat Treating

If you do not properly understand how the blade cools, it is the most common reason the edge of the blade will warp when heat treating. When you grind the piece thinner, or even as thin as the blade tip, you will warp the metal when you move on to the heat treatment phase. When you hollow grind a piece, you can remove more metal than necessary, which will cause it to cool faster than the edge of the blade, resulting in warping.

7. Burning the Steel When Heating

Many beginner bladesmiths do not even realize it is possible to burn steel until they remove their work from the flame, and it crumbles before ever making contact with the hammer. While the actual temperature that the steel will begin to burn varies depending on the steel you use, a clear indication that you have burned your work is if you see sparks flying off the metal prior to placing it on the anvil.

8. Treating Every Steel the Same When Heat Treating

A common mistake for beginners is treating every steel they use the same way. Each steel is different, and when it comes to heat treating, it will need to be processed uniquely. When heat treating steel, you must understand the process necessary for forging that steel, the appropriate temperature it needs to reach, what quenching oil to use, what temperature the quenching oil needs to be at, how to cool the steel properly, and what steps you need to take to create a well-polished piece.

9. Not Maintaining Control over the Steel When Heat Treating

When it comes to heating the metal, you need to be aware of how you place the work in the fire. If you simply just toss it in and then take it out when you think it might be ready, you will find that one side of the work is the appropriate temperature and the other is not. When it comes to heat treating, maintaining control of the work while heating and quenching requires patience and a steady hand. Many beginners quickly remove the metal from the fire and dunk it right into the quenching oil. As a result, what many will experience is a crack or damaged blade. When you quickly place the metal in the quenching oil, you may cause parts of the blade to cool faster than it is possible for the steel to withstand. You need to slowly dip the work into the quenching oil so the cooling process does not happen so suddenly and you reduce the risk of damage to the steel.

10. Overheating the Blade While Grinding After Heat Treating

There is a great deal of debate on whether or not one should grind after heat-treating or if all grinding should be done prior to heat treating. While it is an argument that can go back and forth forever, if you are grinding after heat treating, you need to be aware of the temperature of the metal. Grinding after heat treating can cause the steel to warp if you are trying to rush through the process, especially when using a belt grinder. If you slow down the speed and take your time, then you will not raise the heat and should have no issues with the blade tip or edge deforming.

CHAPTER 9: SHARPENING
MISTAKES

Sharpening your blade is an integral part of bladesmithing. It can also be one of the more complex steps and
processes. A Master bladesmith has spent countless
hours mastering just this one step. Sharpening the blade
does not just impact the blades ability to cut through a
variety of material, it also plays a vital role in the dura-
bility of the edge of the blade. When you are just starting
out in bladesmithing, sharpening is one process you really want to perfect your skills in. Just be sure to avoid
the following mistakes as you do.

1. Not Starting with the Right Grit

It is common knowledge that when sharpening a blade,
you would often start with coarse grit and then move on
to less coarse grit as the sharpening process progresses.
While this is the correct approach to sharpening, the biggest mistake that many beginners make is moving on
from coarse grit to fine grit too soon. This is especially
true when working with a very dull edge. You want to
establish a burr before you move on from using your
coarse grit to a finer grit. If you move on before you have
established a burr, you will find yourself in the sharpening phase for a great deal longer.

Another issue is thinking the blade you are already working with is sharp enough to start off with a finer grit. It is wiser to at least give the blade a few passes with a coarse grit paper to ensure the burr has been established before jumping right into a fine-grit sanding.

2. Using a Dull Knife to Learn How to Properly Sharpen a Knife

When you want to learn how to properly sharpen a knife, it makes sense to pick up a dull knife and begin to work the edges until they form a sharp edge. While this logically seems like the best way to learn, it often results in shaving off a great deal of the knife's edge, leaving you with very little to actually sharpen.

Choosing a knife that has a slightly sharp edge, on the other hand, allows you to learn how to properly hold the tools required for sharpening and see how the edge of the blade changes as you proceed. Using a sharp knife and then transforming the blade to a razor-sharp edge will teach you how to angle the blade properly and show you exactly what a razor-sharp edge is supposed to look like. Using a dull knife may result in a sharp blade but it will oftentimes leave you unable to sharpen the edge so that it is razor-sharp.

3. Thinking the Latest Sharpening Tools and Equipment Will Cut Time and Effort

There are plenty of gadgets you can purchase that can speed up any process of a bladesmithing project. As you develop your skills and learn more about the craft, these fancy gadgets and gizmos can save you time and energy. But, one thing they will not allow you to do is to learn the necessary skills and gain the necessary understanding of every step.

This can be especially true in the sharpening process. There are plenty of tools that can assist you in sharpening your blade, but if you lack the basic understanding of how to hold the blade or what a sharp edge is supposed to look like, then the tool won't help you out in the long run. You still need to develop the necessary skills that will allow you to perform the task flawlessly and while tools can be helpful, they can hinder the development of these necessary skills.

4. Not Taping the Blade When Sharpening

If you have sharpened a blade without taping it first, then you should already have an idea as to why this is a big mistake. If you haven't gotten to the sharpening phase yet, then make sure you avoid these mistakes. Taping the blade when you sharpen it reduces the risk of the grit and debris scratching the blade. If the tape begins to wear away when sharpening, you will want to replace it. You don't want your finished blade to be full of scratches.

5. Skipping out on the Hand Sanding Process

Learning to sand by hand gives you a great advantage when you begin knife making. While the process seems to be time-consuming, in the long run, it can save you a great deal of time as you develop your skills. Hand sanding a piece will help you better understand the angle at which you need to hold the knife to get the desired effect. It will allow you to see how the grit of the sandpaper affects the edge of the blade. It will also give you a better understanding of how steel can be manipulated without any fancy machines. Many beginners fail to realize how valuable it is to learn this traditional technique, but those who take the time to master hand sanding are further ahead than those who rely on machines to do the sanding for them.

6. Not Holding the Blade at the Correct Angle

A beginner bladesmith tends to hold the blade at a low angle when they are sharpening, which results in the sharpening process taking up a great deal of your time. If you hold the blade at too high of an angle, the edge of the blade will be too thin and dull. How you hold the blade will make all the difference and getting accustomed to the right angle is only done through practice. As a general rule, you want to keep the blade at an angle that is just slightly lower than the intended edge angle you are going for.

CHAPTER 10: NOT HAVING THE PROPER TRAINING OR KNOWLEDGE

Bladesmithing requires one to develop and have a clear understanding of various techniques. You don't just simply grab a piece of steel, heat it, and bang away at it until you magically have the perfect blade. The steps, process, and skills required to perfect this art form is an ongoing process. If you fail to have the right understanding of basic knowledge of the steps, you will fail to move forward as a respected bladesmith. This chapter covers the most common mistakes beginners make when it comes to learning the trade or the lack of knowledge they gain during the starting phase.

1. Not Fully Understanding the Process of Heat Treating Steel

Heat treating is a crucial step in the knife making process. When done correctly, the result is a strong, durable blade that can withstand a great deal of wear and tear. If done incorrectly, the blade will crack, shatter, and become deformed when dropped or used. There are a number of ways one can go about heat treating a knife and there are some ways you should definitely not go about heat treating. The most common mistakes beginners make when heat treating is either skipping steps or rushing through the process.

Heat treating is a simple process of knowing how to properly control the temperature of the forge and the right quenching oil. The quenching oil is typically a vegetable oil or another type of cooking oil. It must be at the right temperature when being used. The temperature that everything needs to be kept at will depend on the type of steel you are using at your specific location. A general rule of thumb is to heat the quenching oil to about 130 degrees Fahrenheit and the steel should be a nice red color before quenching.

Though the process is quite simple, it takes practice to understand when each step needs to be performed. Not heating the steel to the right temperature evenly will result in a blade that is only halfway hardened.

2. Getting Information from Books, Online, or Magazines That Are Not Credible

The Internet makes it easier than ever to find detailed information about how to make a knife. There are plenty of individuals who probably have little to no experience hammering a blade but are recording a video to teach others how to do it. You always have to be careful about where you gather your information from. There are plenty of forums and websites that can show you the proper way to forge, grind, and walk you through all the steps. There are many more that pretend to know what they are talking about, which can result in injury, wasted time, and no actual value. Books are some of the best places you can gain your information from. Some well-known Master bladesmiths have written detailed introductions and step-by-step guides that will help you better understand the process. Some books you should definitely read to further your understanding of bladesmithing include:

- *$50 Knife Shop* by Wayne Goddard
- *The Complete Bladesmith* by Jim Hrisoulas
- *Bladesmithing: Beginner, Intermediate, and Advanced Guide to Bladesmithing* by Wes Sander

3. Not Watching a Bladesmith Actually Make a Piece in Person or Thinking You Do Not Need Lessons

If you think you can just learn everything you need to know about bladesmithing by going at it yourself, you might be able to forge a few mediocre knives. If you take the time to actually watch and learn how the process is done by a Master bladesmith, you will be further along to make flawless or near-flawless knives. When you watch an experienced bladesmith go through the process of removing the stock, forging, heat treating, grinding, and finishing the blade from start to finish, you gather a wealth of knowledge. You see exactly how they hold the work, their stance, how they set up their workspace, and you also learn a great deal more about the approaches you should take, which cannot always be clearly translated through text. If you think you can figure it all out on your own, then great; but, if you want to truly learn the craft and art of blade-making, then you need to see an expert do it from start to finish.

4. Not Learning or Even Knowing There Are Different Types of Blows You Can Perform

Beginners have the belief that they can just step up to the anvil and begin hammering away. The thought of, if they just turn the work a certain direction then it will eventually bend and begin to form into the shape they desire, is wrong. They will often use the same straight-on striking motion with the hammer and quickly wear themselves out and become frustrated when the piece is not molding the way they intended it to. This is due to a lack of understanding of the types of blows that can be used to shape and manipulate the steel. The most common blows consist of half-faced blows, offsetting blows, and angle blows. Each one is performed for a different result.

Half-faced blows are when you position the spot you intend to strike near the edge of the anvil. When you strike your hammer, it should make contact with the work where half of it is on the anvil and half of it is off the anvil. The result of this strike creates a cut or dip in the work to help form a distinguished tip or endpoint of the blade. These blows are what you will use to help form the choil of the blade.

Offsetting blows help give the metal more shape and can add curves to your design. This is done by striking the metal slightly off-center as opposed to directly over the anvil.

The angle blow helps shape the tips of the piece into a point. You not only hold the work at an angle on the anvil, but you also strike the hammer at an angle.

In order to achieve the desired effect, you not only need to know where to position the work on the anvil, but you also need to understand the type of blow you need to use. It takes a great deal of practice and commitment to perfect these blows, but there is no way to shortcut this process. The more effectively you can deliver a blow, the faster you can shape the work and the less time you will need to keep reheating it.

5. Not Knowing What a Leg Vise Is For

There are two common vises a bladesmith uses, though they vary depending on the bladesmiths preference. The first is the bench vise. It is a sturdy vise that can be helpful when filing or twisting your work. They, however, are not meant to take on a great deal of stress. A leg vise, on the other hand, is meant to be used while hammering. It is designed to take the stress from the strikes of a hammer while keeping the work secure and in place. The hammering will not cause damage to the piece and the jaws are designed to allow quick and easy removal of it. A leg vise is valuable when you need to keep your work steadier and deliver accurate hits to achieve the desired shape.

6. Not Knowing What Needs to Be Done When You Are at the Anvil

When you are forging your work, you need to know what steps you need to take when you step up to your anvil. Once you have ensured you have all the tools in your reach, you need to know exactly what strikes or hits you need to make to get the desired effect. You need to have patience when heating the metal and not jump into the striking if the metal has not reached the proper temperature. You need to know where you are going to place the work on the anvil when it is removed from the heat. You also need to know how to turn it and hammer it so that you waste little time. You should never start forging your work if you do not already know what steps you need to take to shape the metal the way you drafted.

7. Not Knowing What to Do with Your Half Ground Knife Blanks

Many beginners simply start the grinding process, feel they have done it enough, throw the knife to the side, and start a new project. Part of the bladesmithing craft is fully finishing the piece you are working on. Even if you are just trying to learn the grinding techniques, you need to follow through each piece to the end. This not only solidifies your commitment to the craft, but it also gets you in the habit of completing a piece from start to finish. When you begin to work on more complex pieces, you will be compelled to fix mistakes and will know what mistakes need to be fixed. If you just grind and move on, you never fully grasp the whole process, and this will lead you to give up on a piece that can be salvaged.

8. There Is No Perfect Time to Start Bladesmithing

You don't need a big shop and fancy tools to start bladesmithing. Many beginners look at those who have done this type of work for years and think they need to have everything they have in order to start. This couldn't be further from the truth. The equipment and tools you see skilled blacksmiths use are after years of contemplating which ones to get. They didn't just run out and buy that equipment. They learned, developed their skills, and then stocked their shop with the tools that suit their preferences. You need to develop the skills before you start acquiring the tools.

9. Thinking You Need to Reinvent the Wheel

Bladesmithing has been around for thousands of years. The techniques and processes have varied little over that time, and this is for good reason. There is no need to jump into bladesmithing thinking you have to do something completely outside the box. While it is recommended that you develop your own techniques, you do not have to completely rework the process. You don't need to come up with a fancy new blade design or embellish on the details. Knifemaking is a simple process that requires a great deal of skill. Mastering your skills should be the focus. You don't need to come up with the next best thing in blade-making; you just need to perfect the skills necessary to make the blade.

10. Not Researching How to Make Specific Knives or Swords Before You Give Them a Try

Jumping right into making a bowie knife or Persian style blade might seem like an easy task once you have gotten the basic skills under your belt. If you fail to do any research on the specific knife you intend to make, you will miss out on the information that will allow you to make the knife properly. Before you decide you want to give a certain style of knife a try, you need to learn about its history, components, and elements. Taking the time to gather this information will help you draw up a design and create a knife that is balanced and well-executed. Every Master bladesmith can identify specific properties about a knife and why those properties are important details to know before making it. They do the research to stay true to the authentic roots of the knife while being able to create a functional and long-lasting blade.

11. Moving on to Using Complex Steel Too Soon

There is a reason everyone advises a beginner bladesmith to practice on 1084 steel before diving into using 01 steel. Each steel is different, and some can be very challenging to manipulate and heat. The 1084 steel is ideal for beginners, especially when learning heat treating because it is a fairly straight forward steel. You can learn to strengthen a number of your bladesmith skills before moving onto more complex steel. If you have completed at least two well-polished knives using the 1084 steel, then you need to keep practicing. Do not just make two knives knowing there are plenty of imperfections and fixes that could and should have been made just to move on to different steel. You will only waste time and money. 1084 steel is also favored by beginners because it is not as expensive and can be easier to come by. As a beginner, you should take advantage of using this simpler steel to fully grasp the process of making a piece from start to finish and understand how to properly use the tools.

12. Not Knowing the Different Properties of Steel

Understanding the different properties that steel possesses will help you gain a deeper understanding of how to properly treat the steel you decide to work with. A Master bladesmith has been able to develop his skills by fully understanding these differences in steel. Steel can be made with a combination of elements, though the main properties consist of iron and carbon. The carbon content is what gives the steel its hardness and strength. Other elements consist of manganese, sulfur, phosphorus, and silicon. The most common steels used in knife making include:

- Carbon steel
- Stainless steel
- Tool steel
- Alloy steel

- Galvanized steel

Take your time to read up and research the components of any steel you work with or plan to work with.

13. Wasting Money on Tools When You Are on a Tight Budget

It cannot be stressed enough, not knowing the tools you actually need to get started with will quickly run you into a hole and hinder the skill set you build. The tools you buy will not catapult you to be a Master bladesmith. If you are on a budget, wasting your money on fancy tools that guarantee to move the process along faster will only hinder the development of the necessary skills needed to become a well-respected bladesmith. It is common thinking that if you have the expensive tools, you will be able to create more pieces in less time and therefore be able to make money faster from the pieces you create.

14. Only Skimming Forums for Information Instead of Reading All the Stickies or Asking Questions

Stickies are a type of well-organized and highly interactive forum for bladesmiths of all levels. As a beginner, it is a wise choice to tag into these forums, ask questions, read through all the information you can find on them, and you will often avoid most, if not all the mistakes that many beginners make when they jump into their first projects. Bladesforum.com has taken a great deal of time to organize their stickies at the very top of the forum page so you can easily go through them one by one. Though this may seem time-consuming and boring, if you want to understand the process, skills, and tools needed to become a Master bladesmith, it is wise to take the time and commit to reading through this information. They have categories that deal with all aspects of knife making and even have forums hosted by Master bladesmiths. The best way to learn is to hear about the process from those who have already gone through it. Reading through the stickies and asking questions on these forums will help you avoid a number of beginner mistakes. These also can be further reviewed to learn about more specific mistakes depending on the type of knife you are working on.

CONCLUSION

So, there you have it, the 101 plus mistakes many beginner bladesmiths make that you now don't have to. Whether you are forging or learning to file still, the mistakes outlined in this book have given you vital tips to avoid disaster when you are first starting out.

Before you close up this book and start forging away, there are two bonus mistakes you should keep in mind. First, mistakes are going to happen. It is part of the process, so do not think that just because you know what not to do now that they aren't going to happen. Don't let the mistakes discourage you. Mistakes are how you will learn to properly hone your skills and develop your own unique technique. While some mistakes are unavoidable, the ones listed in this book can easily be avoided. Don't try to cut corners or skip steps in the process because you think they are unnecessary. When you do, you will be met with imperfect works that will only discourage you.

There are some simple rules when it comes to bladesmithing. Also, safety first. Pay close attention to how the tools you use affect the metal and put in the time! Though simple enough, the last one is the hardest to follow. It takes time to transform an old piece of metal into a beautiful functional knife. It will take even more time if you try to cut corners and make any of the mistakes listed in this book.

Lastly, bladesmithing is not a pretty art form. It takes a lot of work, sweat, and frustration to become a Master bladesmith. Those who earn the title of Master bladesmith have respectfully earned it. It takes years of dedication and time. Before you decide you want to be a bladesmith, you need to truly love the craft. Understand that it is not a trade where you will bring in hundreds of thousands of dollars. The individuals who earn the title of Master bladesmith have done so purely because of their love for the craft. If you find yourself falling out of love with it, you may want to rethink why you began in the first place.

It is not uncommon for many skilled bladesmiths to lose interest as the years pass by when they make knives that don't sell. Many find that their loss of interest is due to the fact that they have modernized the process too much. Fancy tools can help you work faster but they can also take away the appeal that got you interested in the craft in the first place. If you find yourself falling out of love with bladesmithing, it might just be you need to go back to the basics, and when you do, you will want to refer to this book to remember all those silly mistakes you need to avoid.

On that note, fire up your forge and get your hammer ready! You have plenty to work with now!

REFERENCES

Alam, M. (n.d.). Blacksmithing for beginners- The noob guide to start with blacksmithing. Retrieved from https://blacksmithcode.com/blacksmithing-for-beginners-the-noob-guide-to-started-with-blacksmithing/

Bern. (2018, June). Essential safety tips for the beginner blacksmith: a brief look at forge safety. Retrieved from https://learntoforge.com/2018/06/15/essential-safety-tips-for-the-beginner-blacksmith-forge-safety/

Kalif, W. (n.d.). How to forge a knife. Retrieved from http://www.stormthecastle.com/blacksmithing/blacksmithing-a-knife/forging-a-knife-part-1.htm

Mark, P. (n.d.). 11 Common Mistakes Beginner Blacksmiths Should Avoid. Retrieved from https://begintoblacksmith.com/11-common-mistakes-beginner-blacksmiths-should-avoid/

Goddard, W. (2018, March). Knifemaking 101- Read this before you make a knife. Retrieved from https://blademag.com/knifemaking/knifemaking-101-read-this-before-you-make-a-knife

Graigson, C. (2017, January). Top 5 worst mistakes made when sharpening your knife, Retrieved from https://mtknives.net/2017/01/mistakes-sharpening-your-knife/

Grundhauser, E. (2015, April). The 10 Trials of the Master Bladesmith. Retrieved from https://www.at-lasobscura.com/articles/the-ten-trials-of-the-master-bladesmith

Pinkerton, P. (2016, November). Knife making- What you need to know about abrasive belts. Retrieved from https://www.outdoorrevival.com/instant-articles/knife-making-need-know-abrasive-belts.html

Shackleford, S. (2017, April). Can you define a knife choil? Retrieved from https://blademag.com/blog/can-define-knife-choil

www.ingramcontent.com/pod-product-compliance
Lightning Source LLC
Chambersburg PA
CBHW061050220326
41597CB00018BA/2726